Flat Earth Science

By Erik ORION

SUN #1
(+)

Earth #1

Earth #2

ORION.

SUN #2
(+)

The Information Contained in This Ebook is a Big

Poke in The Eye to the Illuminati New World Order!

Compliments of Erik ORION

Background of Author

Background of Author

 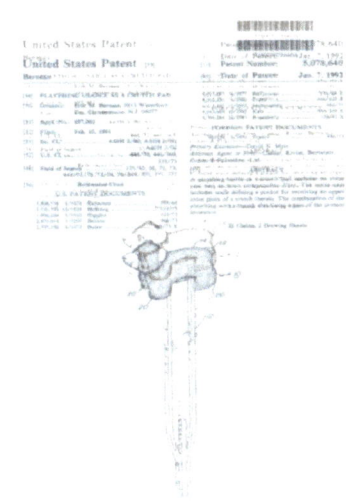 

Erik Berman aka: Erik ORION is 58 years old with a B.A. in Marketing from Stockton State University in Pomona, NJ. He is an author, an inventor with U.S. Patent #5,078,640, a treasure hunter, scuba diver, fisherman, small boat captain, internet webmaster, a licensed real estate broker, a Nikola Tesla historian and an Oak Island theorist / researcher like no other…

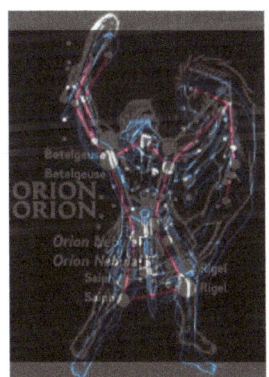

Erik is the world-renowned true conspiracy book author of The Bush Connection book which details the Bush Family's ties to Nazi war criminals & Adolph Hitler's faked suicide as told to him by Commando: Otto Skorzeny. Erik has been the guest speaker on many internet talk shows and radio talk shows to discuss his book The Bush Connection. Erik ORION was recently the guest speaker in a History Channel DVD movie documentary created by Gala Films of Montreal, Canada called U.S. National Parks: Secrets & Legends. Erik wrote the following other amazing books:

1) The Pyramids of Oak Island.
2) Sabotage: The Mystery of Flight 19 is Finally Solved!
3) Tesla Pyramid Power Plants.
4) Communist LGBT Politicians Exposed.
5) Celebrity Death Hoaxes Exposed.

Table of Contents

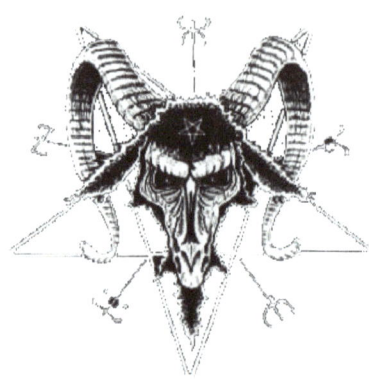

S.A.T.A.N.'s Theory of Our Universe
(**S**cience **A**nd **T**echnology **A**gainst **N**on-Science)

The Christian Science Symbol of the Globe is NOT what our Earth really looks like!

No Globes in Public Schools!

No Globes in Public Schools!

The Globe Represents Christian Non-Science or Nonsense!

Our two Earth's are **<u>NOT</u>** shaped like inflated big blue balls as N.A.S.A. & the Vatican falsely claim.
That's Christian Science Bullshit, Non-Science or Nonsense!

Introduction: <u>**N.A.S.A. & The Vatican are Lying to the World!**</u>

We as human beings have been brainwashed on a "global" scale by NASA, the C.I.A., all the World's Governments, the Vatican & the Freemasons into accepting the Christian "myth" that there is only one Earth & it's shaped like an inflated ball that orbits the sun. I am claiming that the "globe" represents Christian "non-science" or "nonsense" & it should be banned from all American Schools as per the "alleged" doctrine of separation of Church & State.

The globe is a Christian Symbol & should <u>**NOT**</u> be allowed in Public Schools or Public Buildings!

N.A.S.A.'s Big Blue Ball Lie

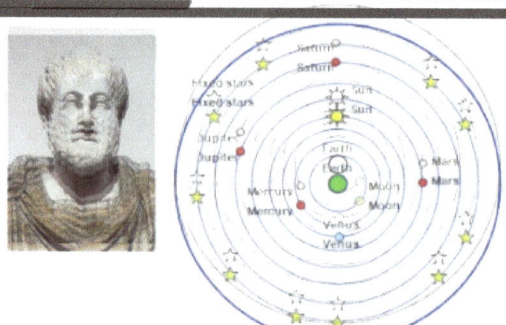

Greek Philosopher Aristotle & Vatican Agent / Christian Scientist
Ptolemy's Geocentric Theory of Our Universe is Obviously Wrong:

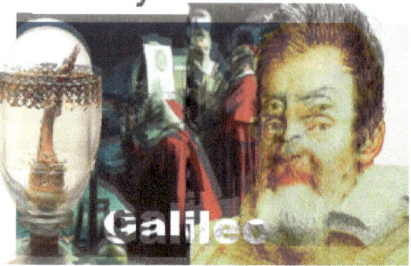

Unfortunately For Mankind, The Vatican Agents / Christian Scientists:
Galileo, Copernicus & Kepler's Accepted Heliocentric Theory of Our
Universe is Also Wrong: It's Not Real Science: It's Christian non-Science
or Nonsense!

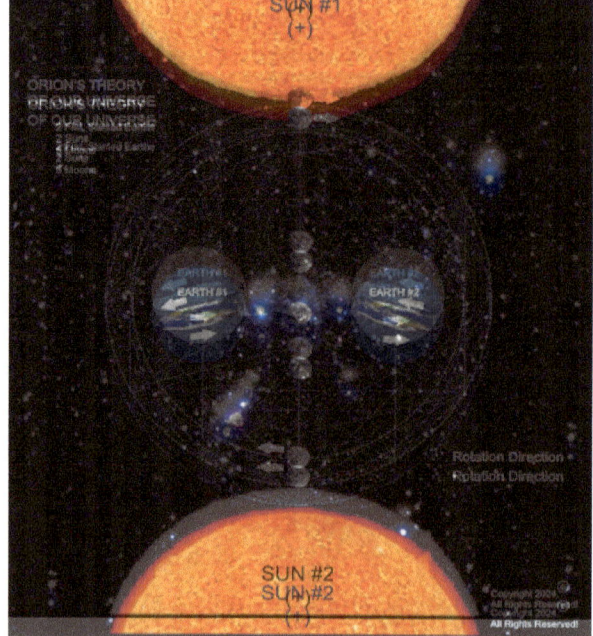

Erik Berman aka:
Erik ORION

American Inventor / Genius: Erik Berman aka: Erik ORION's Amazing Theory
of Our Universe is Based on Sacred Geometry, Mathematics & Real Science:
Not Christian Science Bullshit!

N.A.S.A., the Vatican & the Freemasons claim that our earth it shaped like a big blue inflated ball when in fact, they know it isn't! Their big, blue ball claim is a flat-out lie. It's pure Christian non-science or nonsense! Only high level Freemasons and Jesuit Priests are allowed by NASA to go into space. The general public is supposed to accept and believe whatever they are told about what exists in outer space even if it is a complete lie. NASA faked the alleged moon landings in a film studio in Area-51!

N:A:S:A.'s "version" of our Earth's Solar System is a lie to help promote Christian Science which is "non-science" or nonsense:

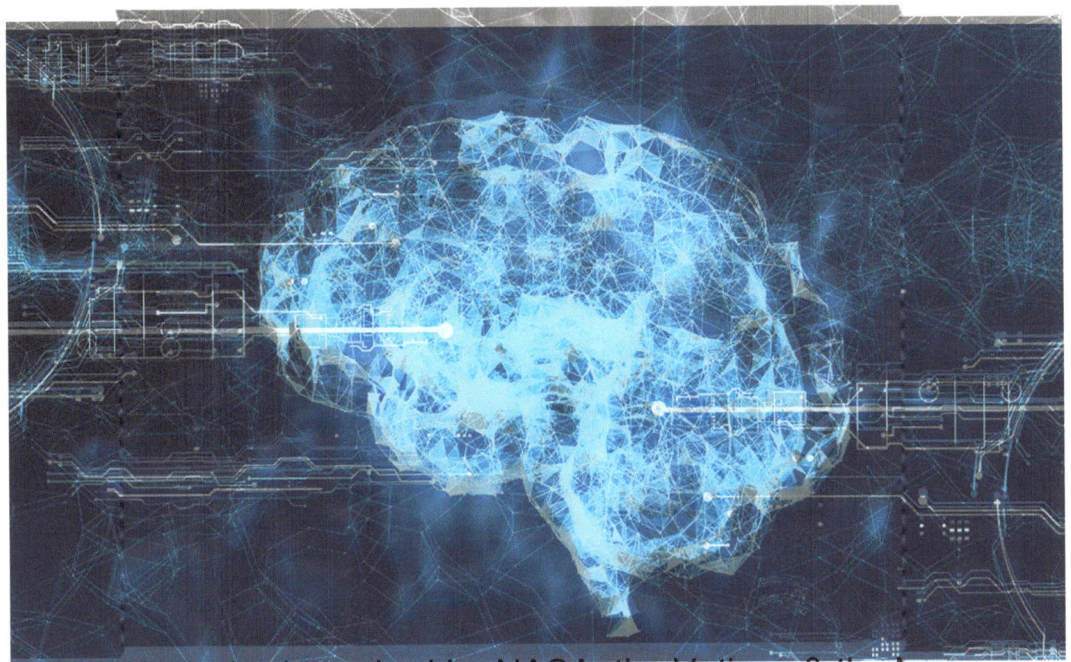

You have been brainwashed by NASA, the Vatican & the News Media.

Chapter 2:   <u>**Flat Earth Christian Science Nonsense**</u>

The "Traditional" Flat Earth Theory as Shown Below is "Scientifically" Impossible.

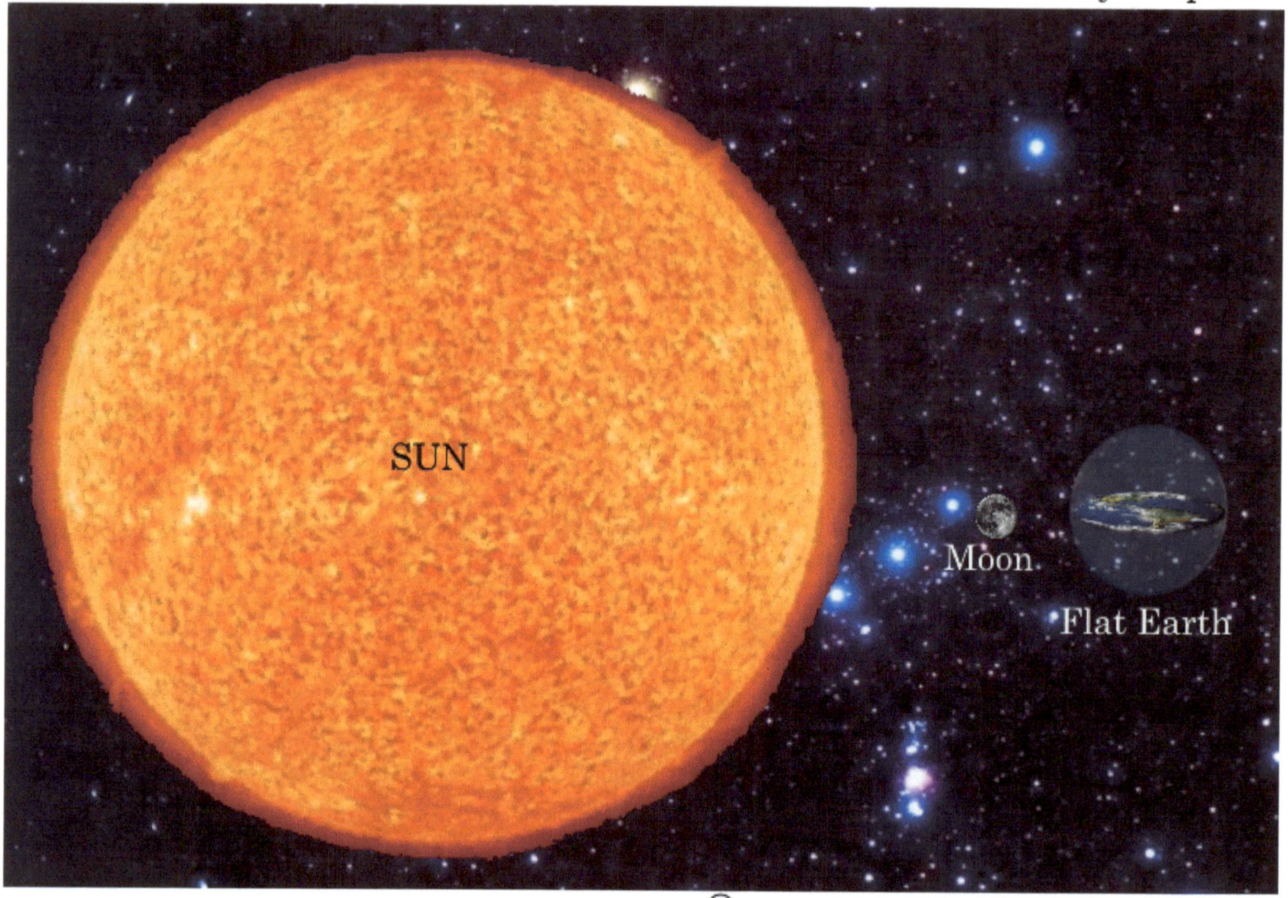

SUN

Moon.

Flat Earth

Many Christians believe the Earth is flat because the Christian "Book of Lies" aka: The Bible says so..

Traditional Flat Earth theorists believe that our Earth looks like this image. Unfortunately, this Flat Earth example is NOT scientifically possible. This is why most Flat Earth Theorists are viewed as "idiots" by our mainstream society.

Chapter 3:   <u>ORION's Slanted Flat Earth Theory of Our Universe</u>

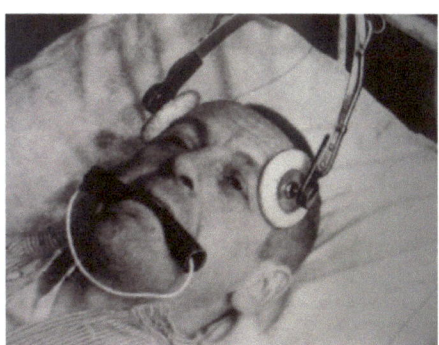

Don't be a brainwashed, blue ball believer any longer!

Hey brainwashed blue ball believers & flat Earth theorists; I am a non-Christian. My theory is based on science & sacred geometry, not Christian non-science or nonsense. Our Planet Earth does <u>not</u> orbit the sun! In fact, there are two planet Earths. Both earths are flat as a pancake, oval disk shaped possibly with hollow centers slanted at 45° angles. However, the Freemasons keep everything level.

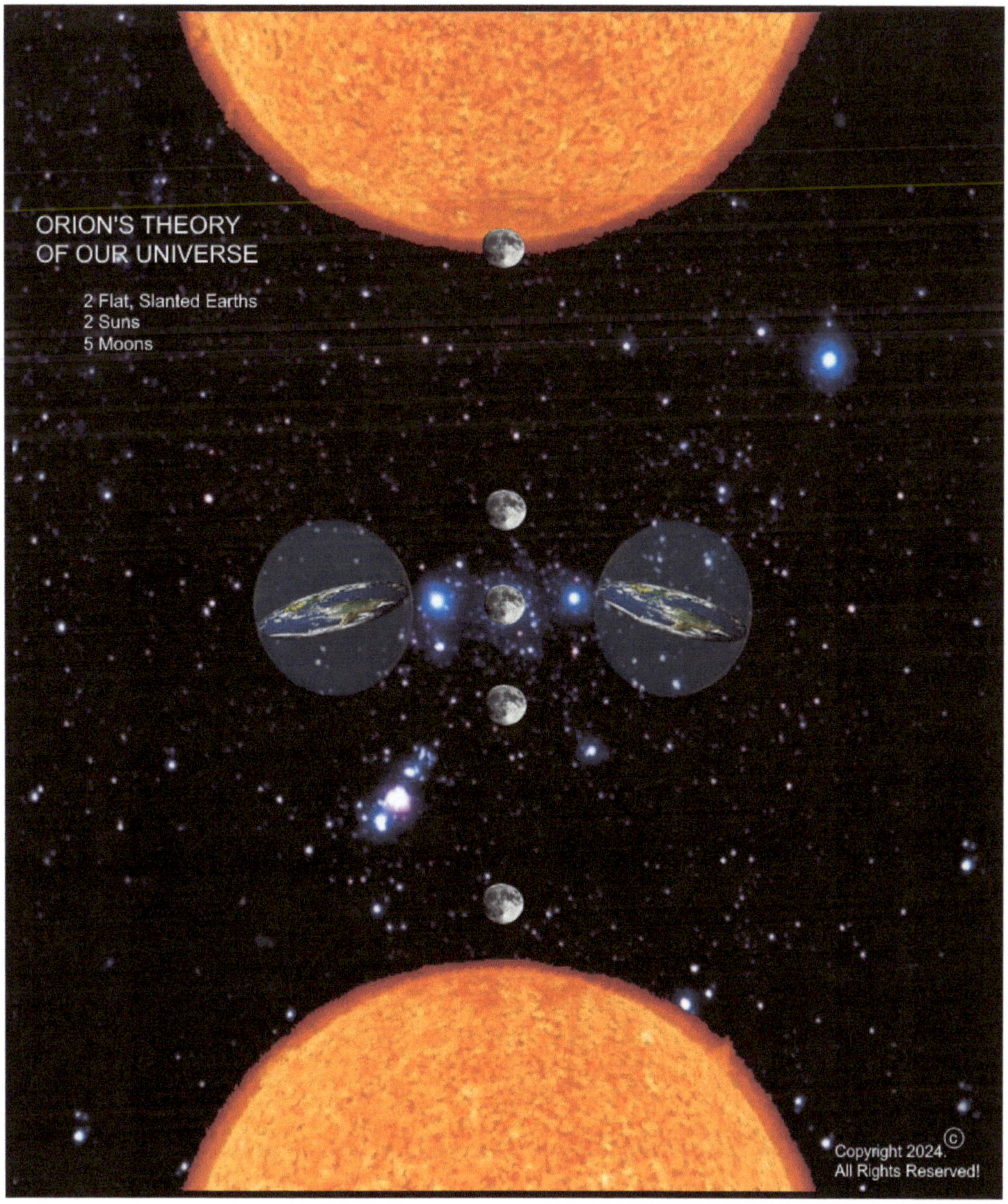

ORION'S THEORY
OF OUR UNIVERSE

2 Flat, Slanted Earths
2 Suns
5 Moons

Earth #2, our "present" Earth is stationary. It does not rotate. Earth # 1, our "past" is stationary too & does not rotate. Both Earths are encased in a clear sphere-like, magnetic field atmospheric bubble / dome with oxygen to support life. I estimate that both planet Earths are 100,000,000 miles away from each other and 100 + or - years in time difference. It's the year 2026 on our present Earth #2 & year 1926 on Earth #1 our past. It's possible for both planet Earths to communicate with each other via digital binary / "Wi-Fi" communications at low frequencies, not high ones as NASA claims. The C.I.A.; N.A.S.A.; NSA & the Vatican keep this scientific fact classified "Top Secret."

ORION's Theory of Our Universe with Labels

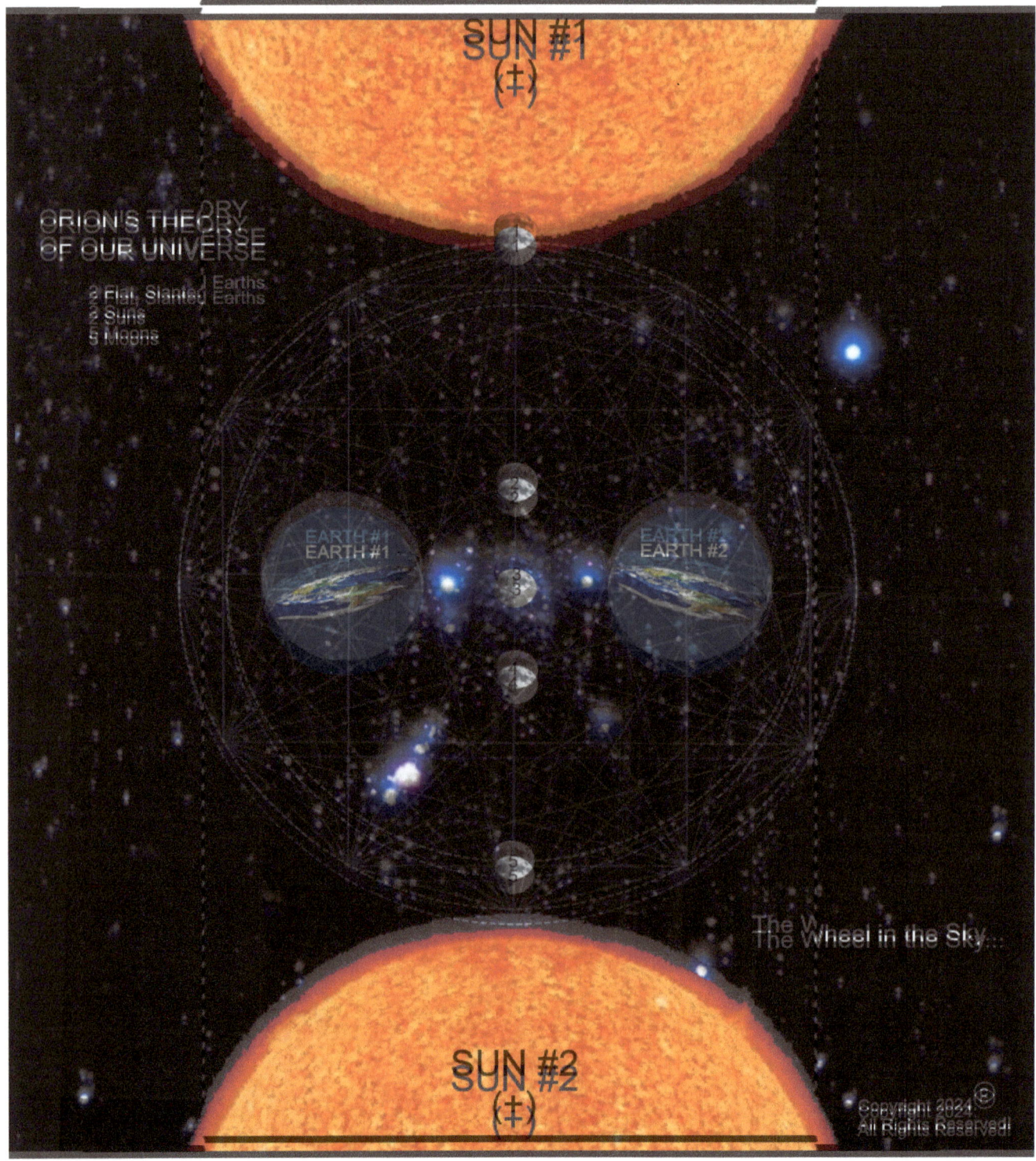

The Universal Clock

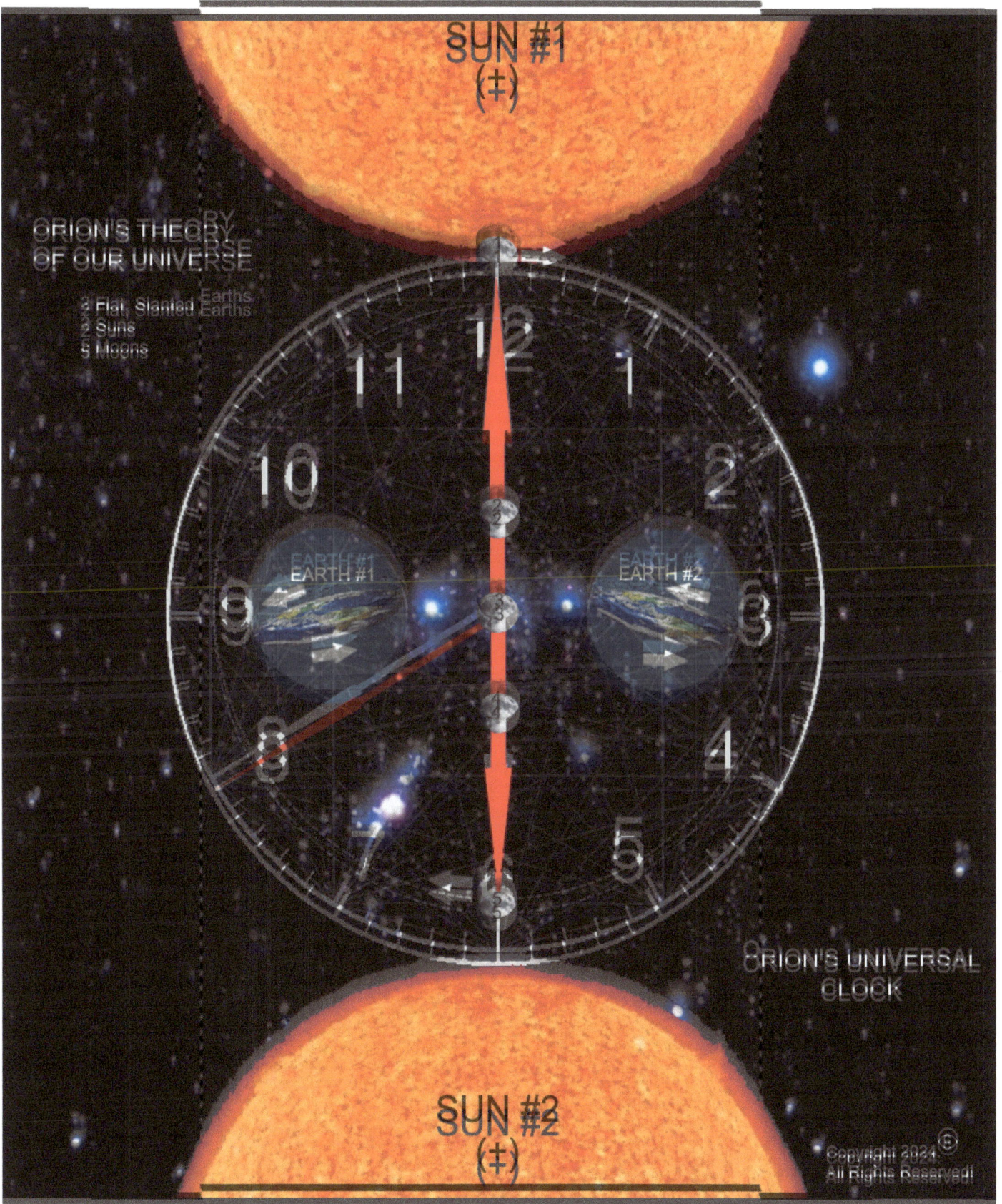

Baphomet

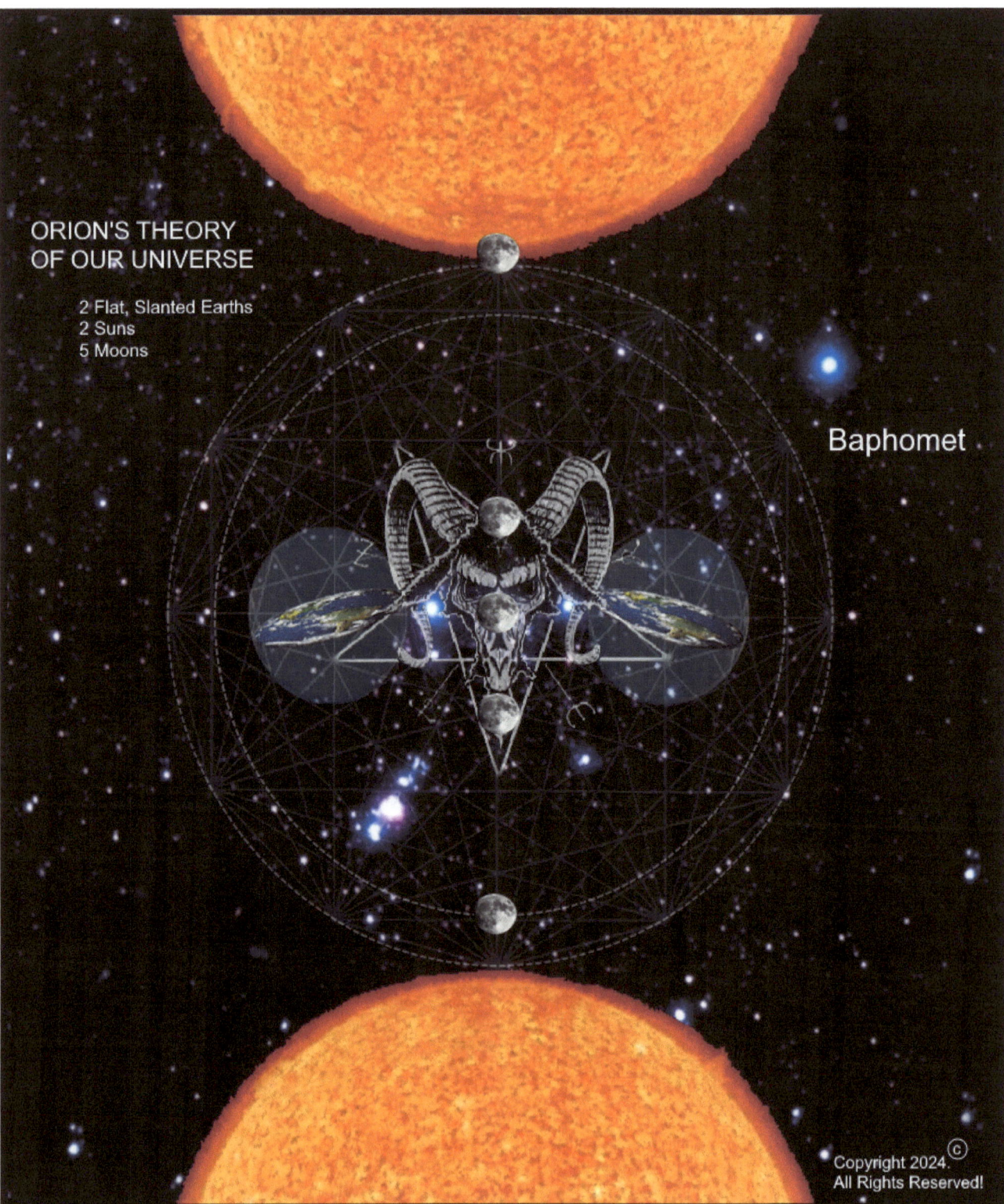

Physics 101; Water always seeks a level surface/plane. (Two 45 Degree slanted, hollow Earths, two Suns, five Moons & four or more solar simulators.) The "Space" above the earth's atmospheric bubble / dome "firmament" might be comprised of water or air & not a gravity-less vacuum because tiny plankton creatures have been found living on the outside of the porthole windows on the International Space Station that only live in water. Furthermore, the Earths might have a hollow tunnel running completely through their center.

Solar Simulators in Orbit

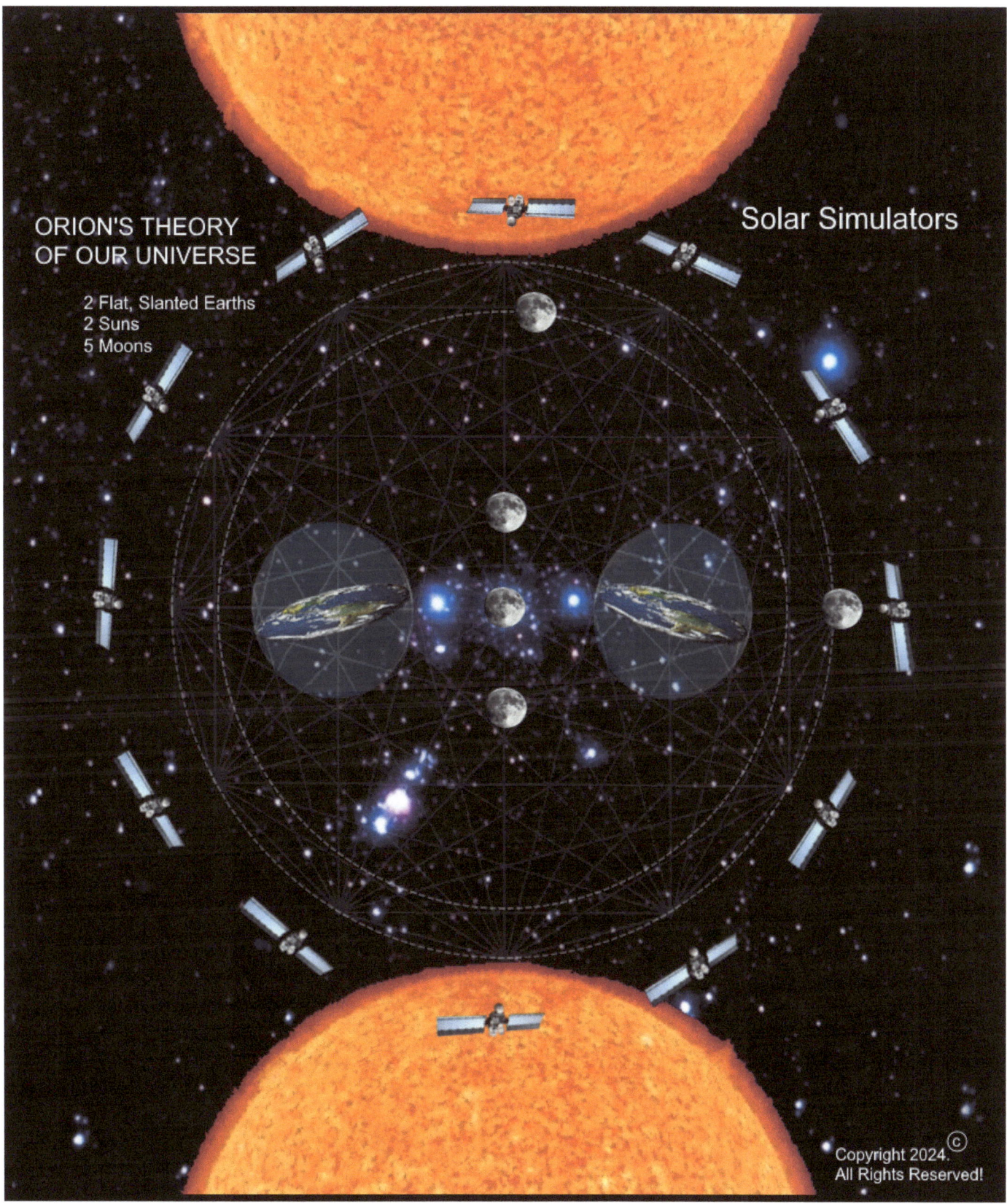

ORION'S THEORY
OF OUR UNIVERSE

2 Flat, Slanted Earths
2 Suns
5 Moons

Solar Simulators

NASA uses patented "solar simulators" to reflect the sun's rays to any location in the world to create "artificial" sunrises and sunsets. Solar simulators are also used manipulate the weather & to reflect the two suns' light energy back into them to help prevent them from losing too much energy & burning out.

Patent # 3,325,238 for a Solar Simulator

June 13, 1967

June 13, 1967

G. GEIER

G. GEIER

SOLAR SIMULATOR

SOLAR SIMULATOR

Filed June 4, 1963

Filed June 4, 1963

3,325,238

3,325,238

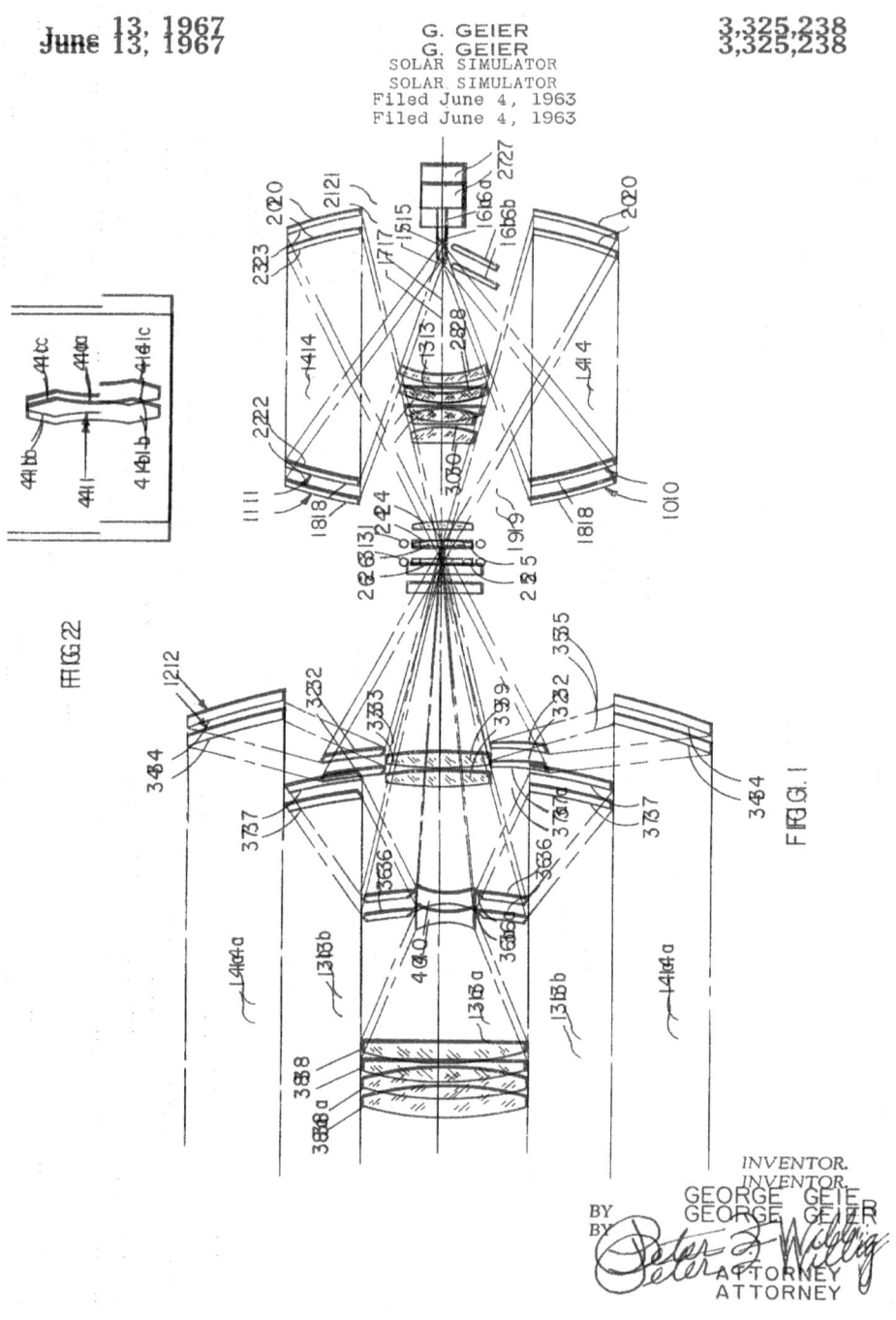

INVENTOR.

INVENTOR.

GEORGE GEIER

GEORGE GEIER

BY

BY

ATTORNEY

ATTORNEY

Solar Eclipse

Solar Eclipse

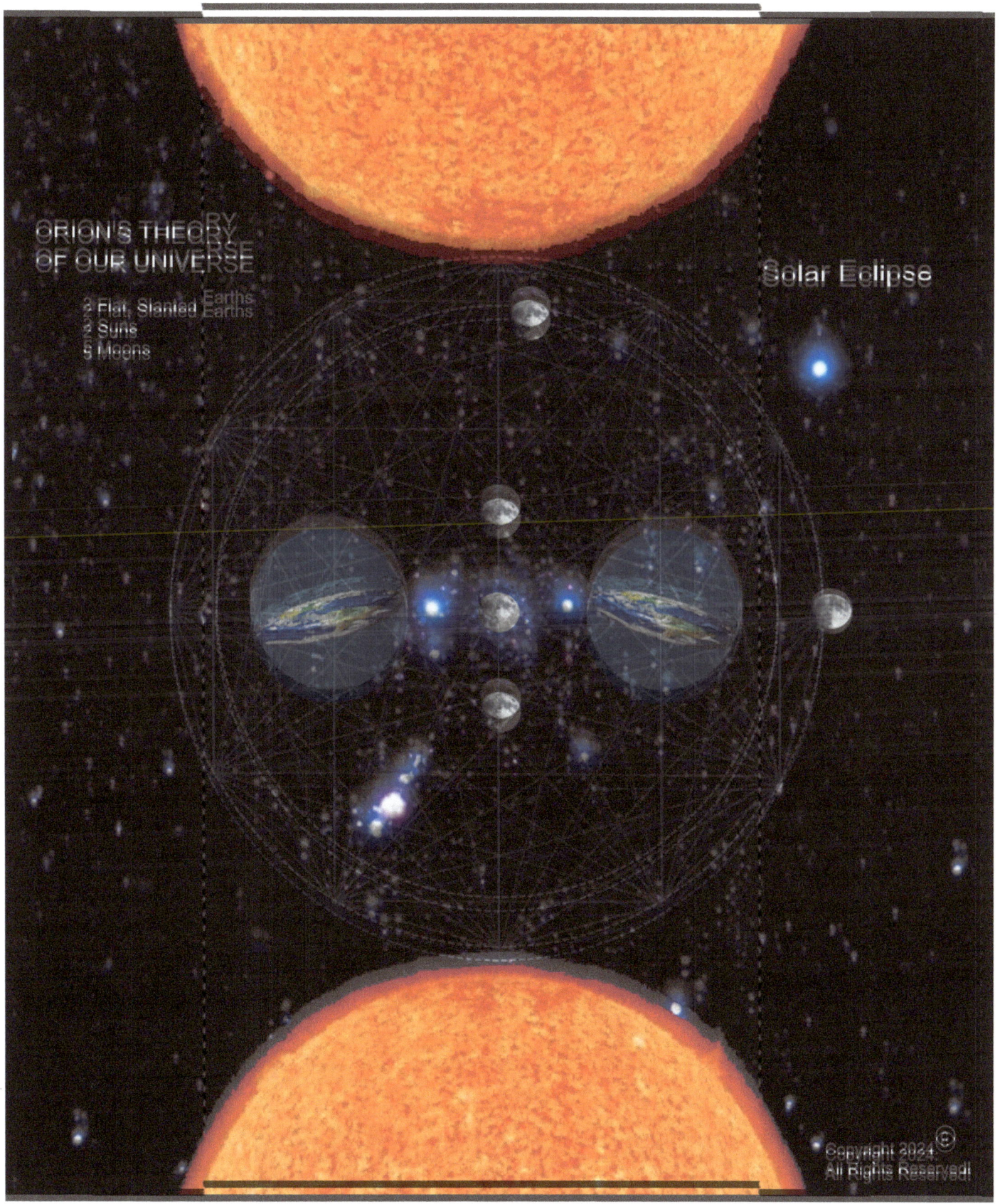

Lunar Eclipse

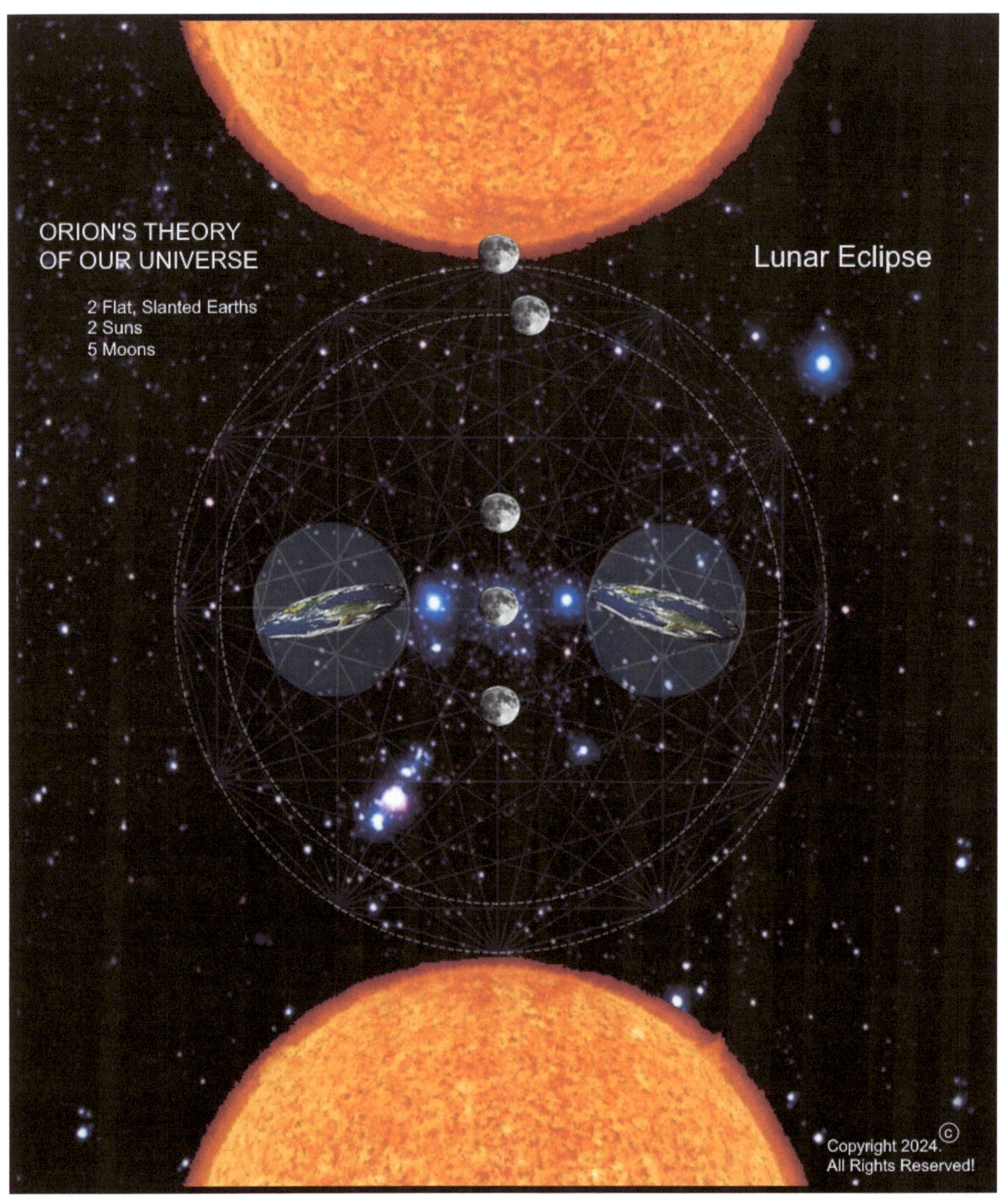

ORION'S THEORY
OF OUR UNIVERSE

2 Flat, Slanted Earths
2 Suns
5 Moons

Lunar Eclipse

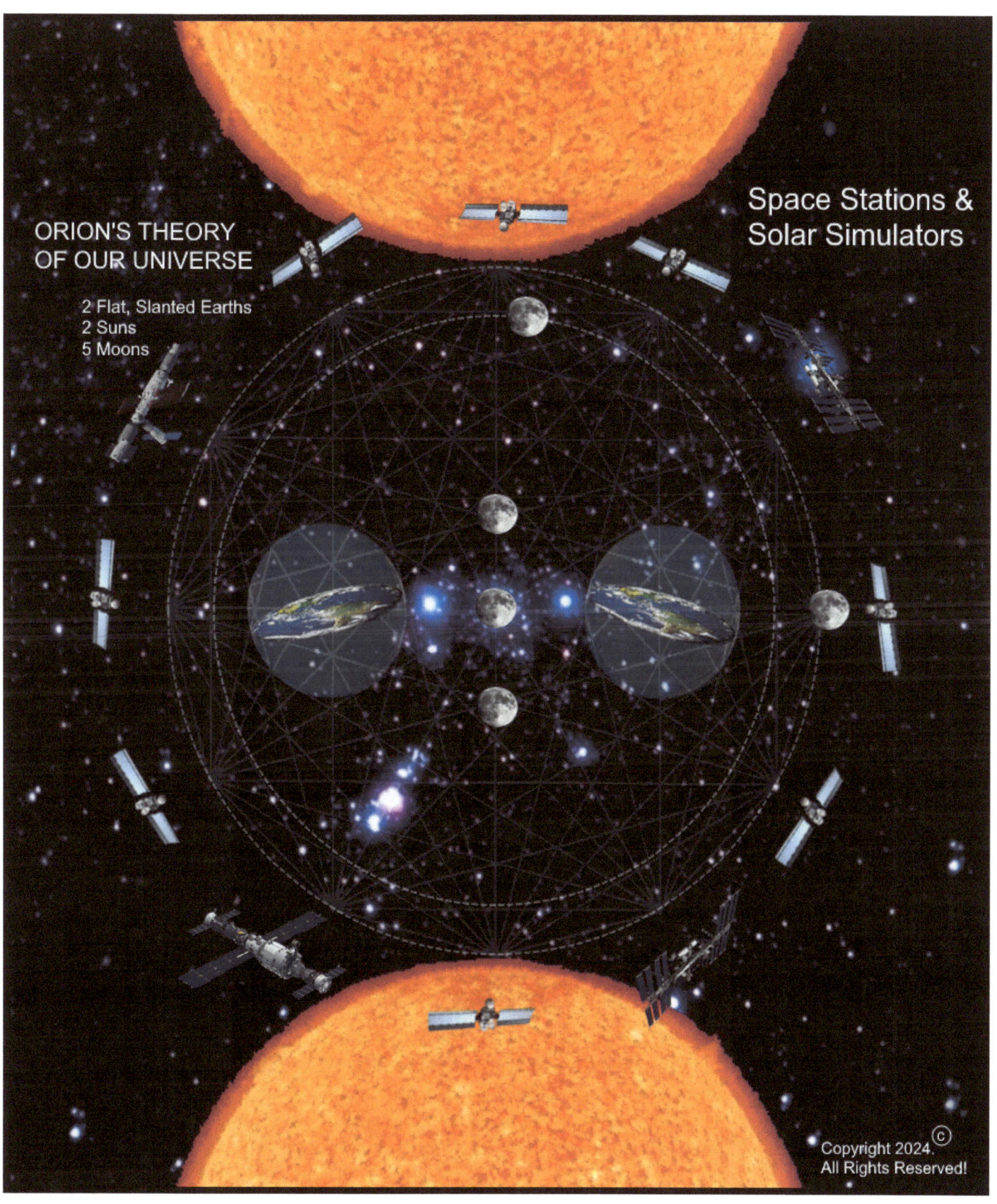

ORION'S THEORY
OF OUR UNIVERSE

2 Flat, Slanted Earths
2 Suns
5 Moons

Space Stations &
Solar Simulators

Chapter 4

Dr. Nikola Tesla: Talking with Planets.

Dr. Nikola Tesla: Talking with Planets.

The idea of communicating with the inhabitants of other worlds is an old one. But for ages it has been regarded merely as a poet's dream, forever unrealizable. And yet, with the invention and perfection of the telescope and the ever-widening knowledge of the heavens, its hold upon our imagination has been increased, and the scientific achievements during the latter part of the nineteenth century, together with the development of the tendency toward the nature ideal of Goethe, have intensified it to such a degree that it seems as if it were destined to become the dominating idea of the century that has just begun. The desire to know something of our neighbors in the immense depths of space does not spring from idle curiosity nor from thirst for knowledge, but from a deeper cause, and it is a feeling firmly rooted in the heart of every human being capable of thinking at all.

Whence, then, does it come? Who knows? Who can assign limits to the subtlety of nature's influences? Perhaps, if we could clearly perceive all the intricate mechanism of the glorious spectacle that is continually unfolding before us, and could, also, trace this desire to its distant origin, we might find it in the sorrowful vibrations of the earth which began when it parted from its celestial parent.

But in this age of reason it is not astonishing to find persons who scoff at the very thought of effecting communication with a planet. First of all, the argument is made that there is only a small probability of other planets being inhabited at all. This argument has never appealed to me. In the solar system, there seem to be only two planets — Venus and Mars — capable of sustaining life such as ours; but this does not mean that there might not be on all of them some other forms of life. Chemical processes may be maintained without the aid of oxygen, and it is still a question whether chemical processes are absolutely necessary to the sustenance of organized beings. My idea is that the development of life must lead to forms of existence that will be possible without nourishment and which will not be shackled by consequent limitations. Why should a living being not be able to obtain all the energy it needs for the performance of its life-functions from the environment, instead of through consumption of food, and transforming, by a complicated process, the energy of chemical combinations into life-sustaining energy?

If there were such beings on one of the planets we should know next to nothing about them. Nor is it necessary to go so far in our assumptions, for we can readily conceive that, in the same degree as the atmosphere diminishes in density, moisture disappears and the planet freezes up, organic life might also undergo corresponding modifications, leading finally to forms which, according to our present ideas of life, are impossible. I will readily admit, of course, that if there should be a sudden catastrophe of any kind all life processes might be arrested; but if the change, no matter how great, should be gradual, and occupied ages, so that the ultimate results could be intelligently foreseen, I cannot but think that reasoning beings would still find means of existence. They would adapt themselves to their constantly changing environment. So I think it quite possible that in a frozen planet, such as our moon is supposed to be, intelligent beings may still dwell, in its interior, if not on its surface.

SIGNALING AT 100,000,000 MILES!

Then it is contended that it is beyond human power and ingenuity to convey signals to the almost inconceivable distances of fifty million or one hundred million miles. This might have been a valid argument formerly. It is not so now. Most of those who are enthusiastic upon the subject of interplanetary communication have reposed their faith in the light-ray as the best possible medium of such communication. True, waves of light, owing to their immense rapidity of succession, can penetrate space more readily than waves less rapid, but a simple consideration will show that by their means an exchange of signals between this earth and its companions in the solar system is, at least now, impossible. By way of illustration, let us suppose that a square mile of the earth's surface — the smallest area that might possibly be within reach of the best telescopic vision of other worlds — were covered with incandescent lamps, packed closely together, so as to form, when illuminated, a continuous sheet of light. It would require not less than one hundred million horsepower to light this area of lamps, and this is many times the amount of motive power now in the service of man throughout the world.

But with the novel means, proposed by myself, I can readily demonstrate that, with an expenditure not exceeding two thousand horsepower, signals can be transmitted to a planet such as Mars with as much exactness and certitude as we now send messages by wire from New York to Philadelphia. These means are the result of long-continued experiment and gradual improvement. Some ten years ago, I recognized the fact that to convey electric currents to a distance it was not at all necessary to employ a return wire, but that any amount of energy might be transmitted by using a single wire. I illustrated this principle by numerous experiments, which, at that time, excited considerable attention among scientific men.

This being practically demonstrated, my next step was to use the earth itself as the medium for conducting the currents, thus dispensing with wires and all other artificial conductors. So I was led to the development of a system of energy transmission and of telegraphy without the use of wires, which I described in 1893. The difficulties I encountered at first in the transmission of currents through the earth were very great. At that time I had at hand only ordinary apparatus, which I found to be ineffective, and I concentrated my attention immediately upon perfecting machines for this special purpose. This work consumed a number of years, but I finally vanquished all difficulties and succeeded in producing a machine which, to explain its operation in plain language, resembled a pump in its action, drawing electricity from the earth and driving it back into the same at an enormous rate, thus creating ripples or disturbances which, spreading through the earth as through a wire, could be detected at great distances by carefully attuned receiving circuits. In this manner I was able to transmit to a distance, not only feeble effects for the purposes of signaling, but considerable amounts of energy, and later discoveries I made convinced me that I shall ultimately succeed in conveying power without wires, for industrial purposes, with high economy, and to any distance, however great.

EXPERIMENTS IN COLORADO

To develop these inventions further, I went to Colorado in 1899, where I continued my investigations along these and other lines, one of which, in particular, I now consider of even greater importance than the transmission of power without wires. I constructed a laboratory in the neighborhood of Pike's Peak. The conditions in the pure air of the Colorado Mountains proved extremely favorable for my experiments, and the results were most gratifying to me. I found that I could not only accomplish more work, physically and mentally, than I could in New York, but that electrical effects and changes were more readily and distinctly perceived. A few years ago it was virtually impossible to produce electrical sparks twenty or thirty foot long, but I produced some more than one hundred feet in length, and this without difficulty. The rates of electrical movement involved in strong induction apparatus had measured but a few hundred horsepower, and I produced electrical movements of rates of one hundred and ten thousand horsepower. Prior to this, only insignificant electrical pressures were obtained, while I have reached fifty million volts.

Tesla's Colorado Springs Laboratory

The accompanying illustrations, with their descriptive titles, taken from an article I wrote for the "Century Magazine," may serve to convey an idea of the results I obtained in the directions indicated. Many persons in my own profession have wondered at them and have asked what I am trying to do. But the time is not far away now when the practical results of my labors will be placed before the world and their influence felt everywhere. One of the immediate consequences will be the transmission of messages without wires, over sea or land, to an immense distance. I have already demonstrated, by crucial tests, the practicability of signaling by my system from one to any other point of the globe, no matter how remote, and I shall soon convert the disbelievers.

I have every reason for congratulating myself that throughout these experiments, many of which were exceedingly delicate and hazardous, neither myself nor any of my assistants received any injury. When working with these powerful electrical oscillations the most extraordinary phenomena take place at times. Owing to some interference of the oscillations, veritable balls of fire are apt to leap out to a great distance, and if anyone were within or near their paths, he would be instantly destroyed. A machine such as I have used could easily kill, in an instant, three hundred thousand persons. I observed that the strain upon my assistants was telling, and some of them could not endure the extreme tension of the nerves. But these perils are now entirely overcome, and the operation of such apparatus, however powerful, involves no risk whatever.

As I was improving my machines for the production of intense electrical actions, I was also perfecting the means for observing feeble effects. One of the most interesting results, and also one of great practical importance, was the development of certain contrivances for indicating at a distance of many hundred miles an approaching storm, its direction, speed and distance travelled. These appliances are likely to be valuable in future meteorological observations and surveying, and will lend themselves particularly to many naval uses.

It was in carrying on this work that for the first time I discovered those mysterious effects which have elicited such unusual interest. I had perfected the apparatus referred to so far that from my laboratory in the Colorado mountains I could feel the pulse of the globe, as it were, noting every electrical change that occurred within a radius of eleven hundred miles.

TERRIFIED BY SUCCESS

I can never forget the first sensations I experienced when it dawned upon me that I had observed something possibly of incalculable consequences to mankind. I felt as though I were present at the birth of a new knowledge or the revelation of a great truth. Even now, at times, I can vividly recall the incident, and see my apparatus as though it were actually before me. My first observations positively terrified me, as there was present in them something mysterious, not to say supernatural, and I was alone in my laboratory at night; but at that time the idea of these disturbances being intelligently controlled signals did not yet present itself to me.

The changes I noted were taking place periodically, and with such a clear suggestion of number and order that they were not traceable to any cause then known to me. I was familiar, of course, with such electrical disturbances as are produced by the sun, Aurora Borealis and earth currents, and I was as sure as I could be of any fact that these variations were due to none of these causes. The nature of my experiments precluded the possibility of the changes being produced by atmospheric disturbances, as has been rashly asserted by some. It was some time afterward when the thought flashed upon my mind that the disturbances I had observed might be due to an intelligent control. Although I could not decipher their meaning, it was impossible for me to think of them as having been entirely accidental. The feeling is constantly growing on me that I had been the first to hear the greeting of one planet to another. A purpose was behind these electrical signals; and it was with this conviction that I announced to the Red Cross Society, when it asked me to indicate one of the great possible achievements of the next hundred years, that it would probably be the confirmation and interpretation of this planetary challenge to us. Since my return to New York more urgent work has consumed all my attention; but I have never ceased to think of those experiences and of the observations made in Colorado. I am constantly endeavoring to improve and perfect my apparatus, and just as soon as practicable I shall again take up the thread of my investigations at the point where I have been forced to lay it down for a time.

COMMUNICATING WITH THE MARTIANS

At the present stage of progress, there would be no insurmountable obstacle in constructing a machine capable of conveying a message to Mars, nor would there be any great difficulty in recording signals transmitted to us by the inhabitants of that planet, if they be skilled electricians. Communication once established, even in the simplest way, as by a mere interchange of numbers, the progress toward more intelligible communication would be rapid. Absolute certitude as to the receipt and interchange of messages would be reached as soon as we could respond with the number "four," say, in reply to the signal "one, two, three." The Martians, or the inhabitants of whatever planet had signaled to us, would understand at once that we had caught their message across the gulf of space and had sent back a response. To convey a knowledge of form by such means is, while very difficult, not impossible, and I have already found a way of doing it. What a tremendous stir this would make in the world! How soon will it come? For that it will sometime be accomplished must be clear to every thoughtful being.

Something, at least, science has gained. But I hope that it will also be demonstrated soon that in my experiments in the West I was not merely beholding a vision, but had caught sight of a great and profound truth.

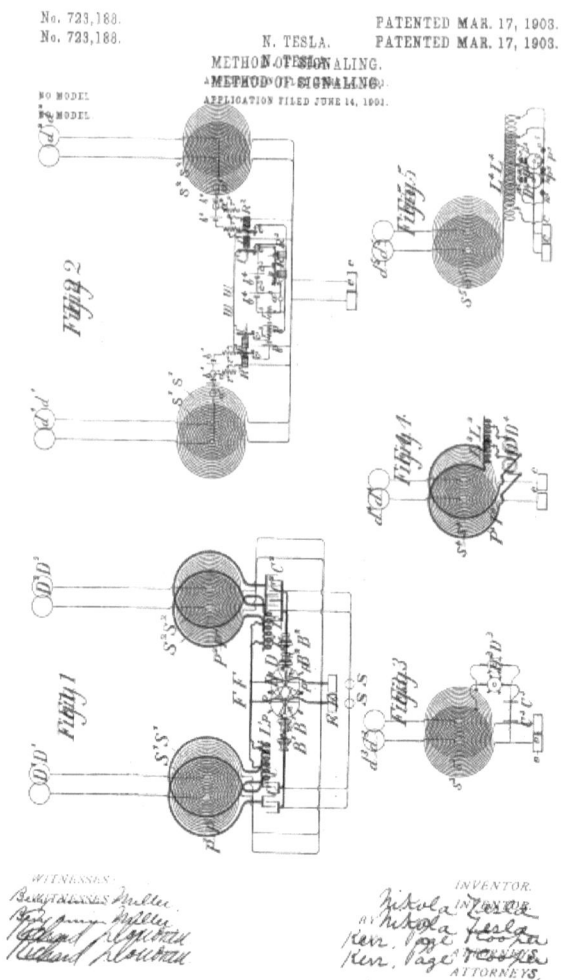

No. 723,188.
No. 723,188.

N. TESLA.

METHO OF SIGN ALING.
METHOD OF SIGNALING.

NO MODEL
NO MODEL

PATENTED MAR. 17, 1903.
PATENTED MAR. 17, 1903.

APPLICATION FILED JUNE 14, 1901.

You don't have to believe me but I'm claiming to have built a Tesla Transmitter in accordance to Tesla Patent #723,188. Furthermore, I'm claiming to be the first person ever to successfully send low frequency, digital binary written & sound messages, photos & videos to another planet. The first message; "1::2::3::" was received by Dr. Nikola Tesla on Planet Earth #1; our past in the year 1899 while Tesla was working at his Colorado Springs Laboratory.

I am also claiming to be the first person to use "telepathy" to communicate to Nikola Tesla Planet Earth #1 via my Tesla Transmitter. Sir William Crookes was there with Tesla when he received my messages. Crookes states as fact that "telepathy" is possible in the realm of Science. Communicating with intelligent life on another planet, time travel & telepathy are scientifically possible. Nikola Tesla & I, Erik Berman are the first ones to prove it.

:

Tesla's thought transmitter.

Conclusion:

N.A.S.A. & the Vatican are lying to the world about the shape of our Planet Earth & how our Universe really looks. There are in fact two planet Earths. Both earths are flat as a pancake, oval disk shaped possibly with hollow centers slanted at 45° angles. Earth #2 rotates counter clock-wise on its axis & Earth # 1 rotates counter-clockwise. Both Earths are encased in a clear sphere-like, atmospheric bubble / dome with oxygen to support life. There are two Suns & seven Moons in our Universe. I estimate that both planet Earths are 100,000,000 miles away from each other and possibly 100+ or = years in time difference. It is the year 2024 on our present Earth #2 & year 1924 on Earth #1 our past. It's possible for both planet Earths to communicate with each other via digital binary / "Wi-Fi" communications. It's also possible to "time travel" to our "past" Earth within 24 hours via NASA space shuttles. N.A.S.A., the C.I.A., N.S.A. & the Vatican keep this scientific fact classified "Top Secret" while selling the Christian Myth of Jesus to the World.

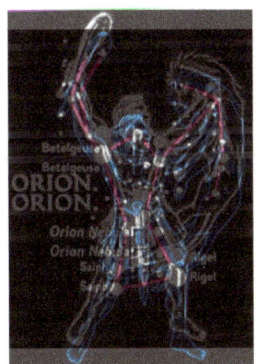

Erik Berman aka: Author: Erik ORION
Self-proclaimed World's Greatest "Unknown" Genius Inventor

The U.S. Government, N.A.S.A. & the Media can contact Erik ORION at:
orionstar123@gmail.com

Episode #4

TESLA'S DEATH RAY: A MURDER DECLASSIFIED ⊕

Discovery

In Hitler's Crosshair's. With Special Guest: Erik ORION; WWII Nazi History / Tesla Subject Expert
ORION is the author of The Bush Connection book which details how Nazis murdered Tesla in 1943!

xfinity

Jack Murphy;
Investigative Journalist

Erik ORION: Author
WWII Historian / Tesla Expert
www.thebushconnection.com

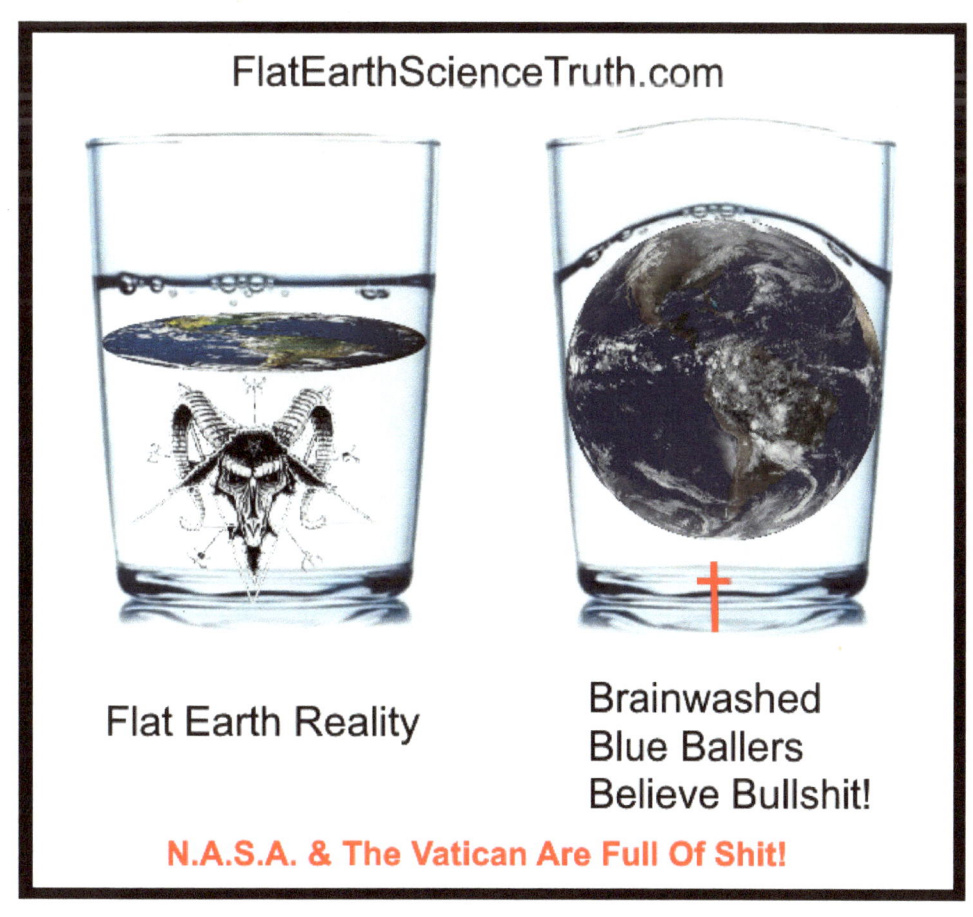

FlatEarthScienceTruth.com

Flat Earth Reality

Brainwashed
Blue Ballers
Believe Bullshit!

N.A.S.A. & The Vatican Are Full Of Shit!

Famous musicians know our Earth(s) are flat. Flat as a pancake apparently.

Famous musicians know our Earth(s) are flat. Flat as a pancake apparently.

The Russians know our Earth(s) are flat:

The Russians know our Earth(s) are flat:

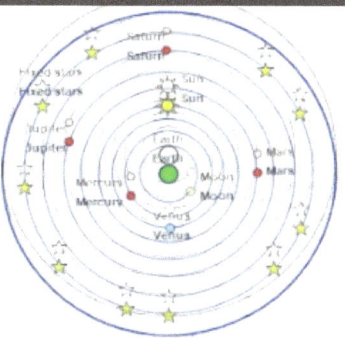
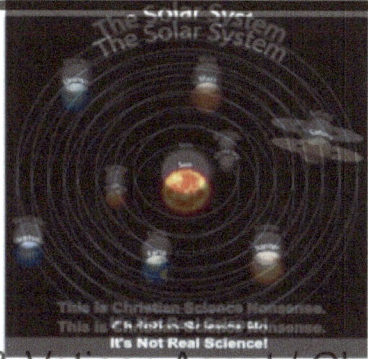

Greek Philosopher Aristotle & Vatican Agent / Christian Scientist Ptolemy's Geocentric Theory of Our Universe is Obviously Wrong.

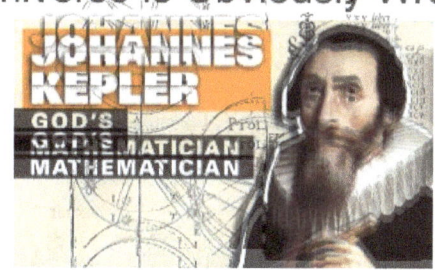

Unfortunately For Mankind; The Vatican Agents / Christian Scientists: Galileo, Copernicus & Kepler's Accepted Heliocentric Theory of Our Universe is Also Wrong: It's Not Real Science; It's Christian non-Science or Nonsense!

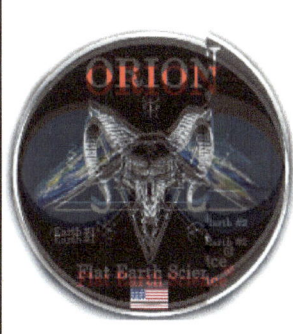

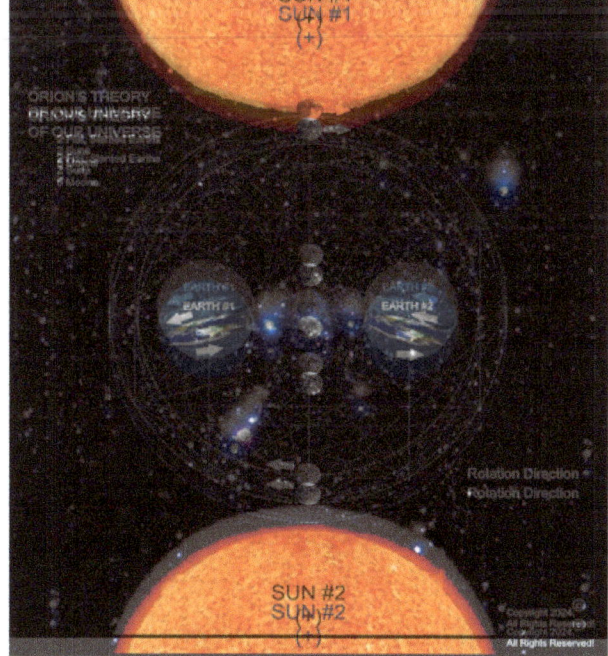

Erik Berman aka: Erik ORION

American Inventor / Genius: Erik Berman aka: Erik ORION's Amazing Theory of Our Universe is Based on Sacred Geometry, Mathematics & Real Science: Not Christian Science Bullshit!

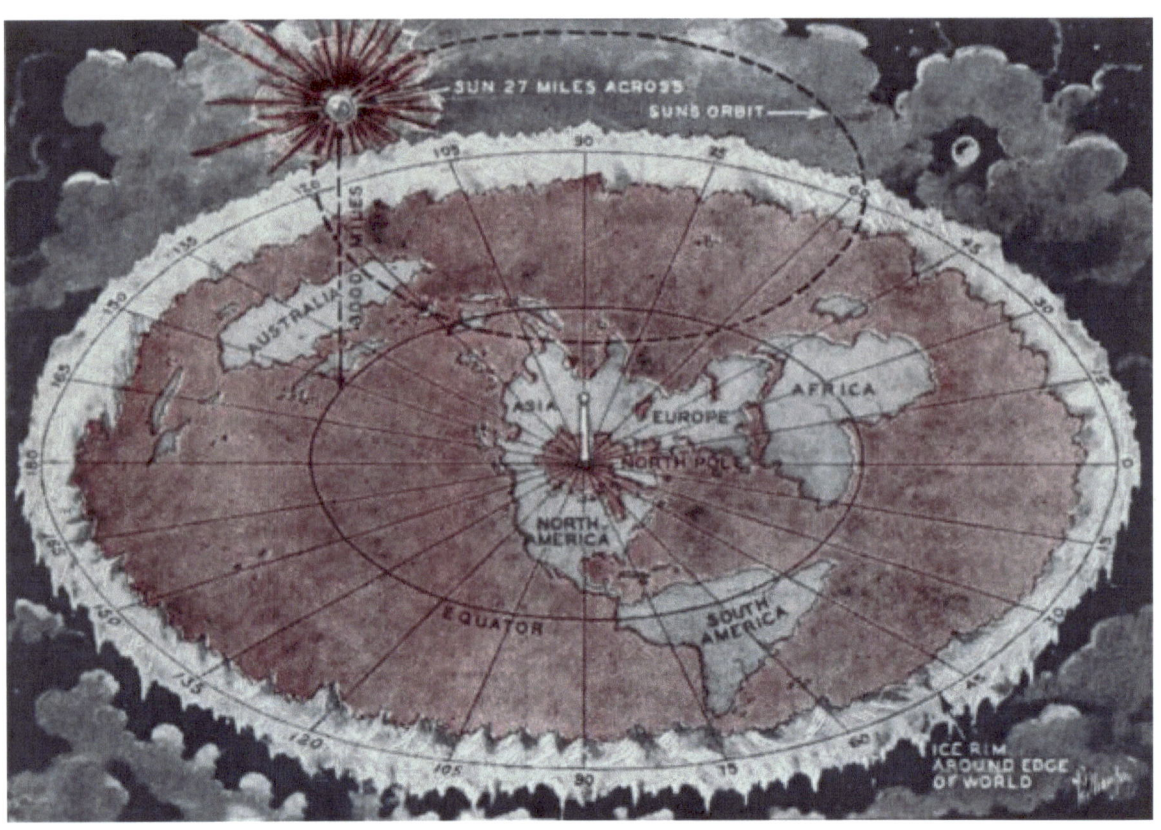

Old Flat Earth Map

Flat Earth Over The Edge
www.flatearthsciencetruth.com
Copyright by Erik ORION 2023. All Rights Reserved!

Hey. Where's The Curve, Bro?

Juno Beach, Florida 2023

www.flatearthsciencetruth.com

The Bush Connection

Flight 19 Sabotage

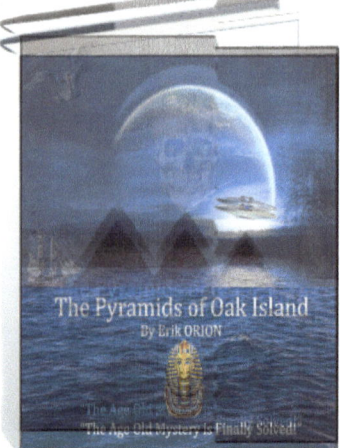

The Pyramids of Oak Island

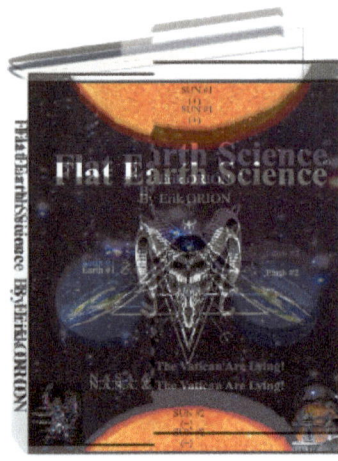

Flat Earth Science

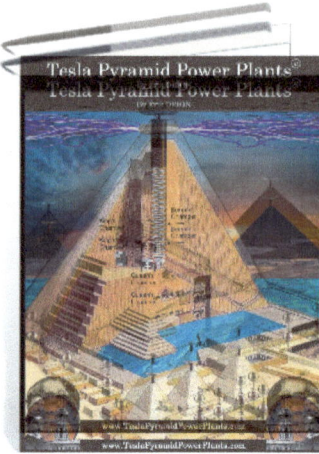

Tesla Pyramid Power Plants

Celebrity Death Hoaxes Exposed

Erik ORION's Amazing Books!